W9-BAK-837

Grasshopper

Karen Hartley,
Chris Macro
and Philip Taylor

Heinemann Library
Des Plaines, Illinois

Published by Heinemann Library,
an imprint of Reed Educational & Professional Publishing,
1350 East Touhy Avenue, Suite 240 West
Des Plaines, IL 60018

Text and cover designed by Celia Floyd
Printed and bound in Hong Kong/China by South China Printing Co. Ltd.

03 02 01 00 99
10 9 8 7 6 5 4 3 2 1

Library of Congress Cataloging-in-Publication Data

Hartley, Karen, 1949-
Grasshopper / Karen Hartley, Chris Macro, and Philip Taylor.
p. cm. – (Bug books)
Includes bibliographical references and index.
Summary: A simple introduction to the physical characteristics, diet, life cycle, predators, habitat, and lifespan of grasshoppers.
ISBN 1-57572-798-6 (lib. bdg.)
1. Grasshoppers—Juvenile literature. [1. Grasshoppers.]
I. Macro, Chris, 1940- . II. Taylor, Philip, 1949- .
III. Title. IV. Series.
QL508.A2H325 1999
595.7'26—dc21 98-42674
CIP
AC

Acknowledgments

The Publishers would like to thank the following for permission to reproduce photographs: Ardea, p. 8; J. Daniels, p. 17; P. Goetgheluck, pp. 5, 11, 13, 14, 26; J. Mason, p. 10; Bruce Coleman Limited/J. Burton, p. 25; W. Cheng Ward, p. 12; M. Fogden, p. 7; H. Reinhard, p.18; K. Taylor, pp. 6, 24; Garden and Wildlife Matters, p. 27; S. Apps, p. 15; K. Gibson, p. 19; Trevor Clifford, pp. 28, 29; NHPA/ S. Dalton, pp. 20, 21, 22; H. and V. Ingen, p.16; Okapia/P. Clay, p. 23; M. Wendler, p. 4; Oxford Scientific Films/L. Crowhurst, p. 9.

Cover photos: Gareth Boden (child); Bruce Coleman Ltd./ J. Brackenbury, (grasshopper).
Illustration: Pennant Illustration/ Alan Fraser, p. 30.

Note to the Reader

Some words are shown in bold, **like this.** You can find out what they mean by looking in the glossary.

Contents

What Are Grasshoppers? 4
What Grasshoppers Look Like 6
How Big Are Grasshoppers? 8
How Grasshoppers Are Born 10
How Grasshoppers Grow 12
What Grasshoppers Eat. 14
Which Animals Eat Grasshoppers? 16
Where Grasshoppers Live 18
How Grasshoppers Move. 20
How Long Grasshoppers Live 22
What Grasshoppers Do. 24
How Are Grasshoppers Special? 26
Thinking About Grasshoppers 28
Bug Map. 30
Glossary. 31
More Books to Read. 31
Index 32

What Are Grasshoppers?

Grasshoppers are **insects**. They have six legs. Most insects have wings. There are many kinds of grasshoppers. They fall into two groups. One group is short-horned. The other is long-horned.

Long-horned grasshoppers have long **feelers**. Katydids and Mormon crickets are long-horned. Short-horned grasshoppers have short feelers. Locusts are short-horned.

What Grasshoppers Look Like

Grasshoppers have long bodies. They are green, brown, or sand-colored. Some have two front wings and two back wings. Some have four wings. Others have no wings.

Their two back legs are long. They are for jumping. Their front legs are shorter. They are for holding food. They use their **feelers** to smell.

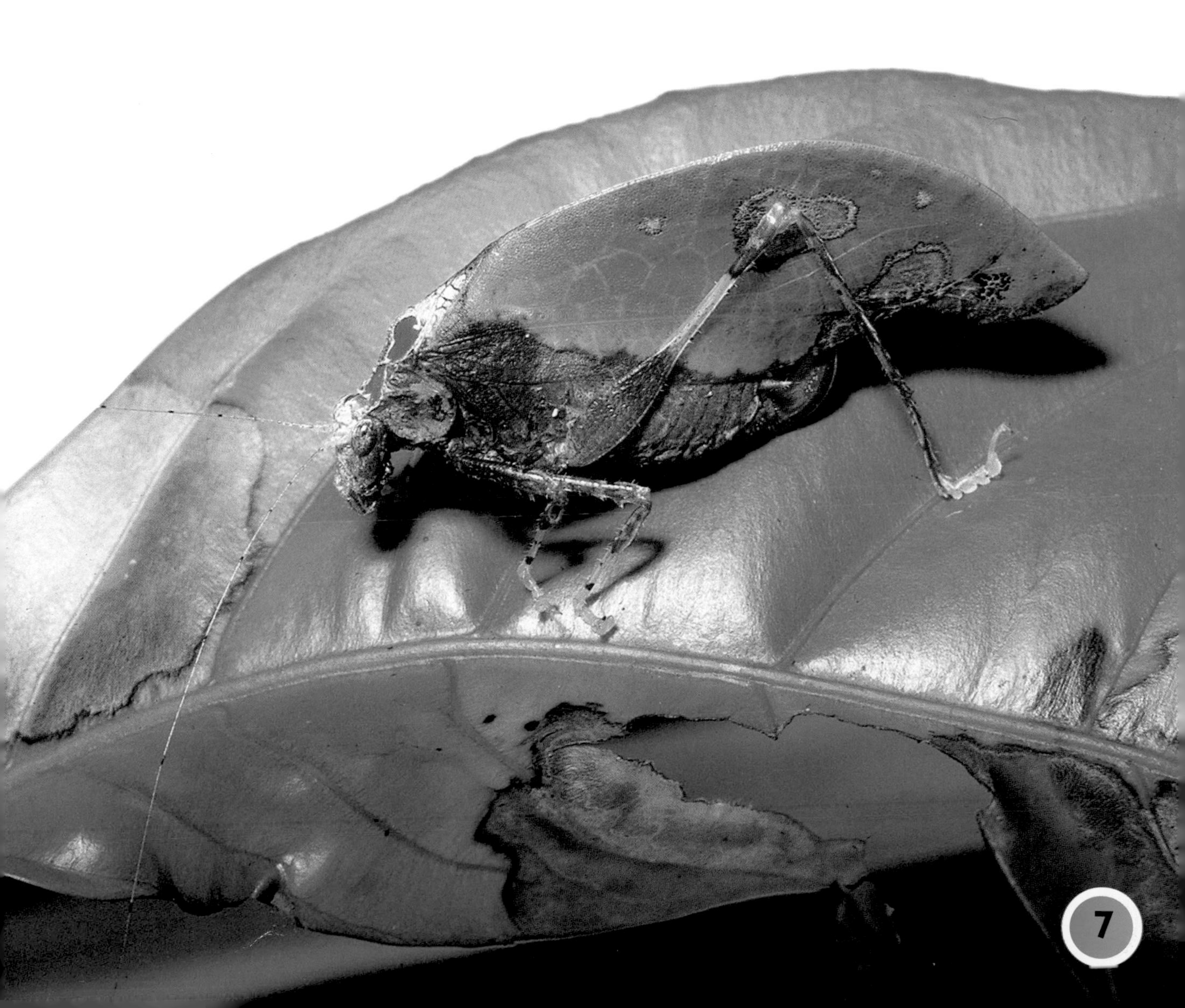

How Big Are Grasshoppers?

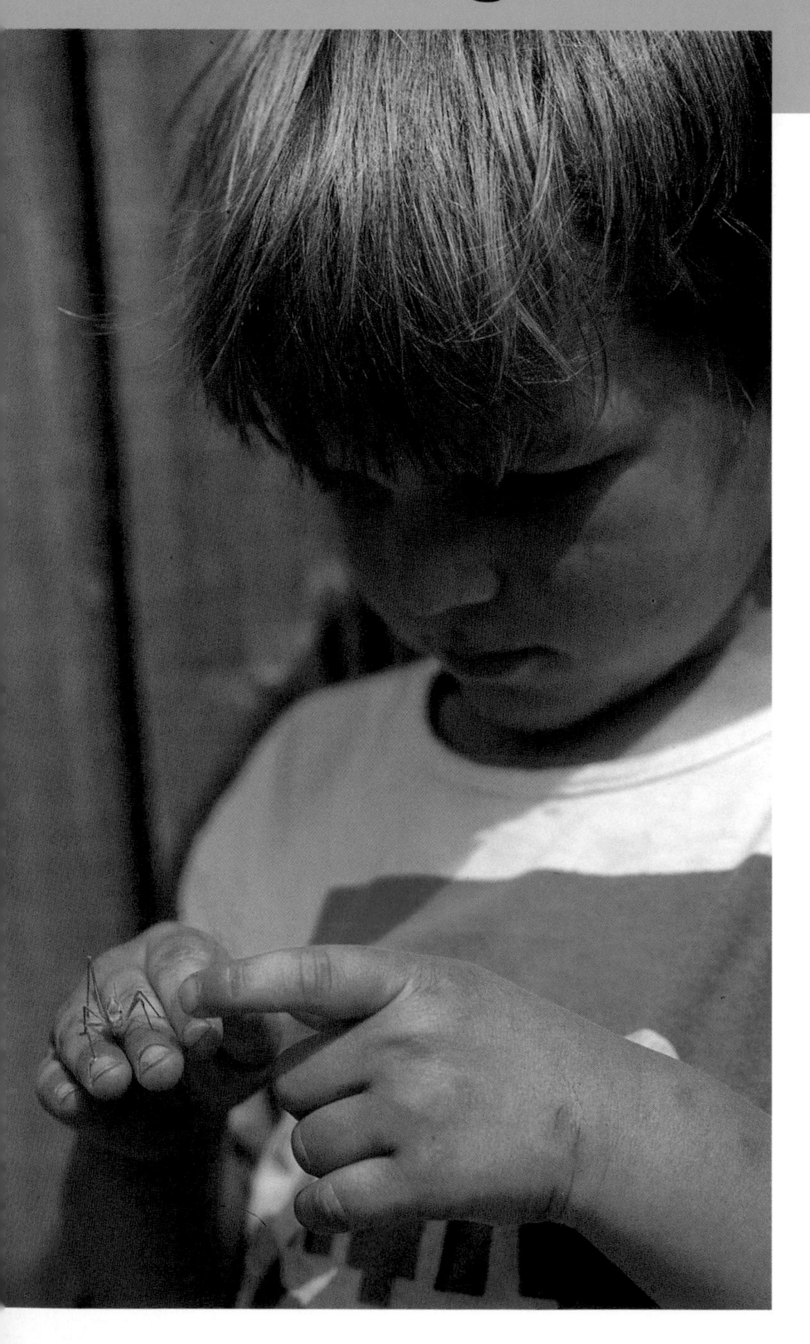

Grasshoppers come in all sizes. Many grasshoppers are about as long as your little finger.

Females are longer than **males**. Look at your little finger. Add another fingernail. That's how much bigger the female grows. Size is the only way to tell males from females.

How Grasshoppers Are Born

The **female** grasshopper digs a hole in the ground. She lays between 2 and 120 eggs at a time. She covers them with a sticky liquid. The liquid gets hard.

The female makes a **pod** that keeps water out. Eggs are laid in late summer and fall. They **hatch** in spring. Babies look like the grown ups without wings.

How Grasshoppers Grow

A grasshopper becomes an adult in 40 to 60 days. To grow, it sheds its skin. When the old skin gets too small, it falls off and gets replaced with new skin.

Grasshoppers **molt** and change their skins five or six times. They molt for the last time when they have grown their wings. Having wings makes it an adult.

What Grasshoppers Eat

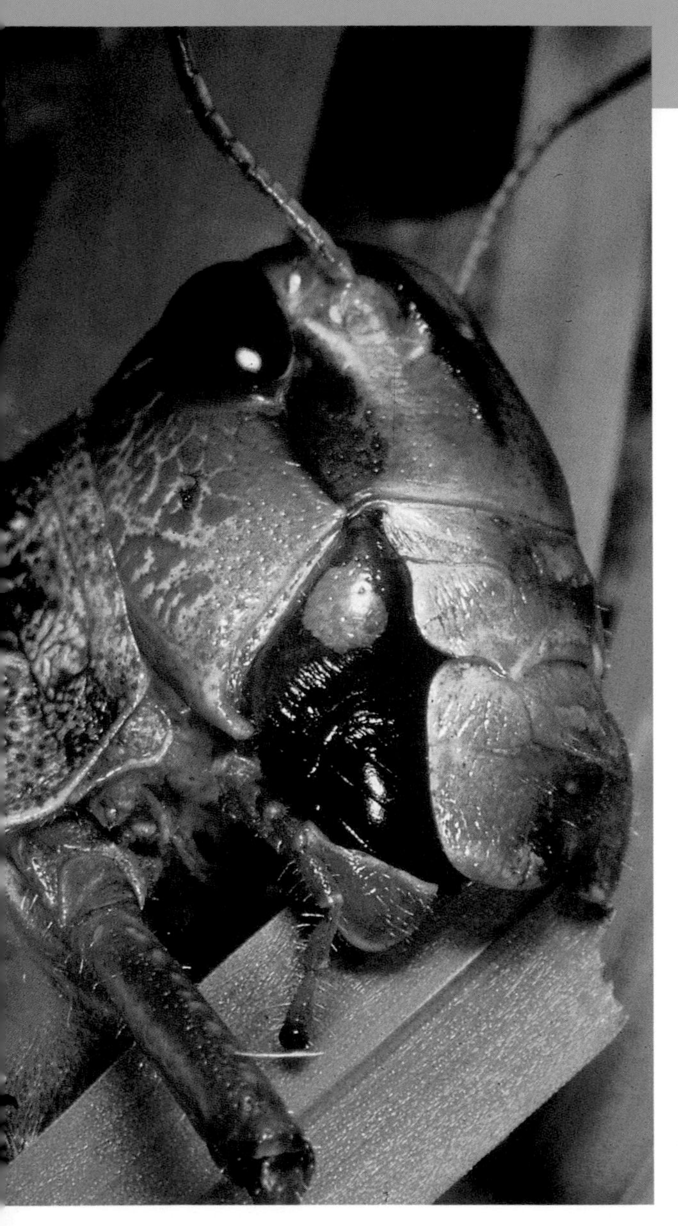

Grasshoppers have two lips and jaws. Their jaws move from side to side when they chew. Short-horned grasshoppers eat plants. Some eat only one kind of plant. Others eat whatever plants they find. **Swarms** of locusts can eat whole fields of plants.

Long-horned grasshoppers eat plants too. But some also eat dead animals. A few kinds eat other **insects.**

Which Animals Eat Grasshoppers?

Many animals eat grasshoppers. Beetles eat them. Spiders eat grasshoppers if they land on their webs. Frogs eat grasshoppers. They catch them with their long tongues. Birds, mice, snakes, and lizards will also eat grasshoppers.

Sometimes grasshoppers can get away. They jump away. They fly away. They hide. Sometimes they bite an animal to get away. Sometimes they spit brown juice on their **enemy**.

Where Grasshoppers Live

Grasshoppers live all over the world except near the North and South Poles. Most grasshoppers live in large fields. Other types live in sand or on cliffs.

Some grasshoppers live in peoples' houses. Some live underground. Other grasshoppers live near beaches or on the banks of streams.

How Grasshoppers Move

Grasshoppers use all their legs to walk. Grasshoppers can jump too. They jump very high. Grasshoppers have very strong muscles in their back legs.

They use these muscles to jump.
They use these muscles to fly. Only grasshoppers with long, thin back wings can fly. Locusts are the best flyers.

How Long Grasshoppers Live

Grasshoppers have short lives. After they **hatch,** they grow. When they are adults they **mate.** They die after they mate and lay eggs. Grasshoppers live from spring to late summer or fall.

Grasshoppers cannot live through a cold winter. But their eggs can! Their eggs are safe and warm underground. They are protected by the **pod.** These grasshopper eggs will hatch in spring.

What Grasshoppers Do

Male grasshoppers sing. But they don't use their mouths! Long-horned grasshoppers rub their front wings together. Short-horned grasshoppers rub their back legs on their front wings.

Each type of grasshopper has its own song. It tells **females** where they are. Grasshoppers sing more if it is sunny and dry. Some sing into the night.

How Are Grasshoppers Special?

How do **female** grasshoppers hear the songs? Grasshoppers have ears. But they don't have them on the side of their heads like you.

Long-horned grasshoppers have ears on their front legs. Short-horned grasshoppers have ears above their back legs. Can you find each grasshopper's ears?

Thinking About Grasshoppers

These children are looking for grasshoppers. They hear them, but they don't see them. They want to watch one for a day or two.

They have a plastic tank with them. How can they make the tank ready for the grasshopper? What will the grasshopper need? Where should they put the tank when they get home?

Bug Map

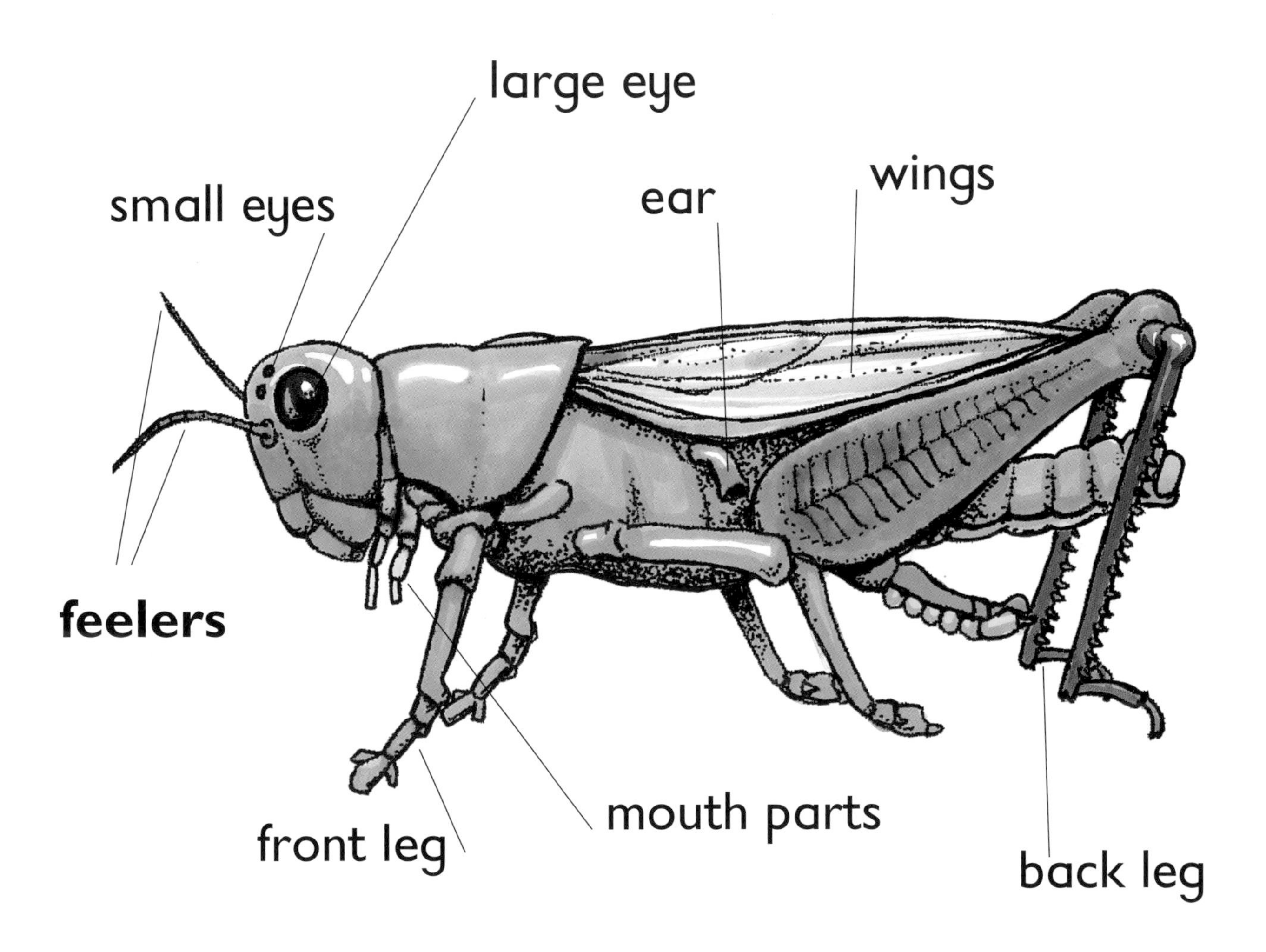
large eye
small eyes
ear
wings
feelers
front leg
mouth parts
back leg

Glossary

enemy one who would do harm

feeler part of an animal's body for sensing touch or smell

female woman or girl

hatch to come out of an egg

insect small animal with six legs, a body that has three parts, and usually having wings

male man or boy

mate to join with another to make babies

molt to shed skin

pod case that holds eggs

swarm large group of insects flying or moving together

More Books to Read

Kalman, Bobbie. *Bugs and Other Insects.* New York: Crabtree Publishing Company, 1994.

Mariner Books Staff. *First Guide to Insects.* New York: Houghton Mifflin Company, 1998.

Index

crickets 5
ears 26, 27
eggs 10, 17, 23
enemies 16, 17
eyes 6
feelers 5, 7
females 9, 22, 25, 26
flying 21
hatching 11, 12
insects 4, 15
jumping 20
Katydids 5
legs 7, 20, 24, 27
locusts 5, 14
males 9, 22, 24, 26
mandibles 14
molting 12, 13
noise 24, 25, 26, 27
pod 10, 11, 23
size 8, 9
swarm 15
wings 6, 11, 13, 21, 24